BAHAMAS PRIMARY

Mathematics Workbook 3

The authors and publishers would like to thank the following members of the Teachers' Panel, who have assisted in the planning, content and development of the books, led by Dr Joan Rolle, Senior Education Officer, Primary School Mathematics, Department of Education:

Deidre Cooper, Catholic Board of Education

Vernita Davis, Ministry of Education Examinations and Assessment Division

LeAnna T. Deveaux-Miller, T.G. Glover Professional Development and Research School

Joelynn Stubbs, C.W. Sawyer Primary School

Dyontalee Turnquest Rolle, Eva Hilton Primary School

Karen Morrison and Daphne Paizee

HODDER
EDUCATION
AN HACHETTE UK COMPANY

Contents

Topic 1 Getting Ready

Number Work

1 Fill in the missing numbers on each number line.

a

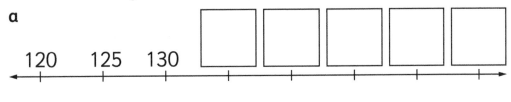

120 125 130

b

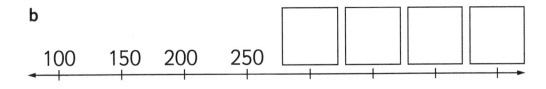

100 150 200 250

c

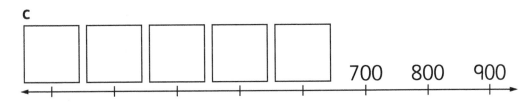

700 800 900

d

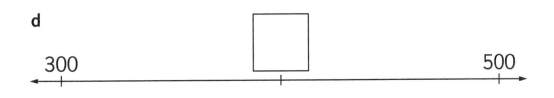

300 500

2 Circle the largest number in each set. Underline the smallest number.

a 234 456 245 623

b 112 300 102 210

3 Look at this number line.

400 500

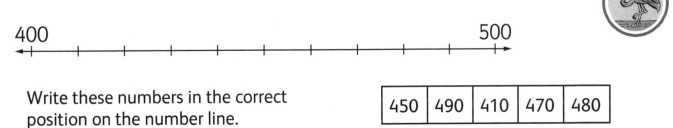

Write these numbers in the correct position on the number line.

| 450 | 490 | 410 | 470 | 480 |

1

A Mixed Bag

1 Sandra counted the shapes in a pattern.

Circles	Squares	Triangles	Rectangles
12	7	11	16

Show this information on the graph.

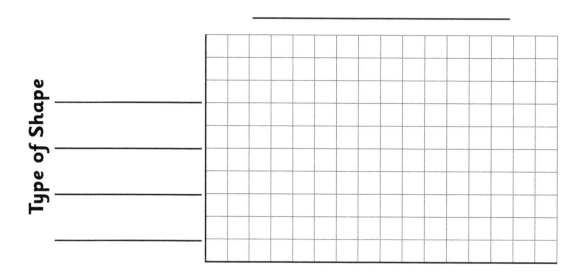

2 Draw a line to divide each shape into halves. Colour one half.

Draw another line to show fourths. Colour one fourth. Use a different colour.

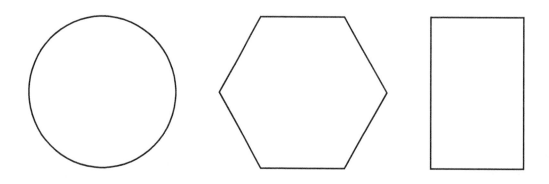

Topic 2 Numbers in Our World

Ordinal Numbers

1 Label the runners in this race with these ordinal numbers:

1st, 2nd, 3rd, 4th, 5th.

2 Write the following numbers as ordinal numbers.

24,   33,   6,   30,   12,   45

_____

3 There are five boys standing in a line. Their names are Sean, Mario, Jesse, Micah and Brad. Fill in the names under the pictures. Here are some clues:

Sean is not second. Mario is not first. Jesse is third in line. Brad is not 5th. Jesse is behind Micah. Brad arrived before Sean.

_____

4 Mrs Adderley has a very busy month. Read the calendar and answer the questions.

January						
Mon	Tues	Wed	Thurs	Fri	Sat	Sun
		1 Book club meeting	2	3 Tennis	4	5
6	7 School meeting	8	9	10	11	12 Picnic
13	14 Me!	15	16	17 Tennis	18	19
20	21	22	23	24	25	26
27	28	29	30	31		

a Write the dates of the first and last days of January.

_____

b Write the dates of the Wednesdays that fall on odd numbers.

_____

c Write the dates on which Mrs Adderley will play tennis.

_____

d What will Mrs Adderley do on the 12th?

_____

e On which date is Mrs Adderley's birthday?

_____

f Mrs Adderley visits her mother on the 2nd and 4th Saturdays of every month. Circle these dates on the calendar.

g Mrs Adderley is going camping for three days from the Wednesday in the 4th week of the month. Colour these days on the calendar.

Roman Numerals

1 Complete the numbers on this clock.

2 Write down the ordinal numbers that match these Roman numerals.

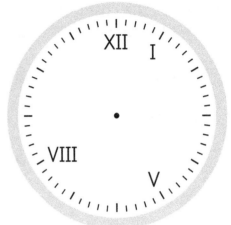

a XIII _____

b XI _____

c XIV _____

d X _____

e XXII _____

f VII _____

3 Write these numbers in Roman numerals.

a 4 _____

b 9 _____

c 13 _____

d 18 _____

e 21 _____

f 25 _____

4 Create a logo for your sports club. The logo should include a Roman number; for example, if your club is 5 years old, you could include that number.

Numbers We Use Every Day

1 Add up the cost of the items on this grocery list. How much money will you have left over from $20.00 if you buy these items?

Apples	$2.30
Ice cream	$2.35
Toothpaste	$1.65
Shampoo	$1.50
Soap	$1.00
Total cost:	
Change:	

2 Make up licence plates with digits that add up to these numbers. The licence plates can have 4 or 5 digits.

a 17

• NASSAU •

•BAHAMAS•

b 15

• NASSAU •

•BAHAMAS•

c 21

• NASSAU •

•BAHAMAS•

d 19

• NASSAU •

•BAHAMAS•

Topic 3 Exploring Patterns

Repeating and Growing Patterns

1 Complete these repeating shape patterns. Use a ruler to draw straight lines. Colour the completed patterns.

a

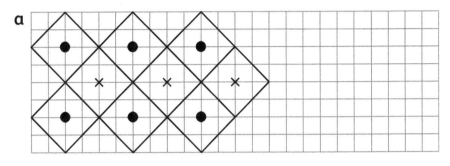

b

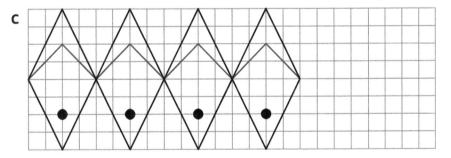

c

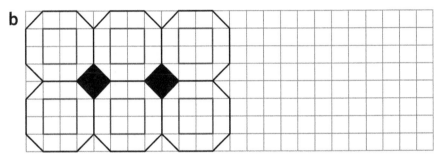

2 Draw a repeating shape pattern of your own here.

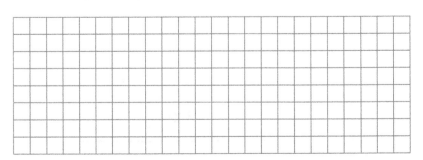

3 Draw the element that comes next in each growing pattern.

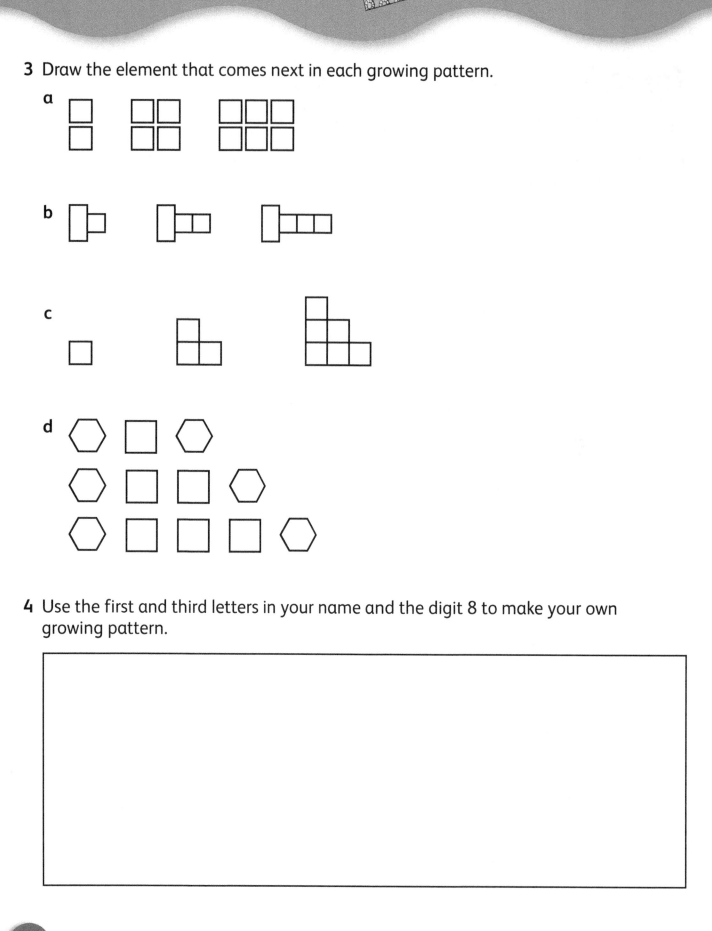

a

b

c

d

4 Use the first and third letters in your name and the digit 8 to make your own growing pattern.

Patterns and Number Lines

1 Use the rule and fill in the missing numbers on the number lines.

a Count forwards by 5s.

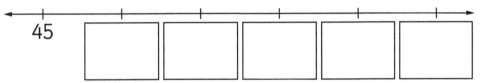

b Add 100 each time.

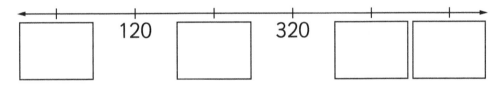

c Count back by 10s.

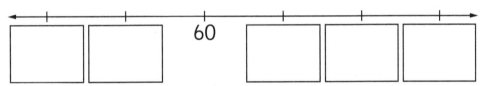

d Subtract 4 each time.

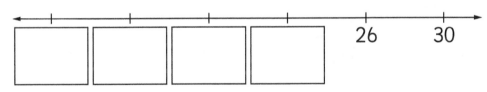

e Add 6 each time.

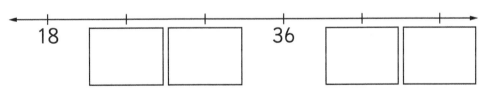

f Add the two previous numbers to get the next number.

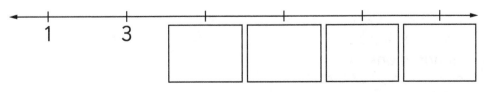

Topic 4 Counting and Place Value

Place Value

1 Colour the shapes to match the number, its name and its expanded notation form. Use the same colour for each set.

| 1 345 | three thousand one hundred twenty-three | 1 000 + 400 + 50 + 3 |

| 3 123 | one thousand four hundred fifty-three | 1 000 + 500 + 50 + 4 |

| 1 453 | one thousand three hundred forty-five | 3 000 + 10 + 2 |

| 3 012 | two thousand ninety-nine | 3 000 + 100 + 20 + 3 |

| 1 554 | three thousand four hundred twelve | 1 000 + 300 + 40 + 5 |

| 2 099 | one thousand five hundred fifty-four | 3 000 + 400 + 10 + 2 |

| 2 601 | three thousand twelve | 2 000 + 90 + 9 |

| 3 412 | two thousand six hundred one | 2 000 + 600 + 1 |

2 Read the information about the number.

Write the digits in the correct places to make a four-digit number that fits the information.

a I am between 1 000 and 1 500.

4, 7, 8, 1

Th	H	T	O

b I am greater than 3 000 and I have 4 tens.

8, 7, 3, 4

Th	H	T	O

c I am between 4 370 and 4 390.

3, 8, 4, 1

Th	H	T	O

d I am between 2 000 and 2 500.

3, 9, 8, 2

Th	H	T	O

e I am greater than 2 780 but smaller than 2 850.

8, 7, 4, 2

Th	H	T	O

f I have 6 hundreds and I am greater than 3 000.

1, 4, 3, 6

Th	H	T	O

g I have no tens and more than five hundreds.

0, 2, 4, 6

Th	H	T	O

h I am between 7 000 and 7 060.

1, 0, 7, 9

Th	H	T	O

Comparing Numbers

1 Start on 4000. Move across the rocks in ascending order. You must jump to the next highest number each time. Draw arrows as you move. Colour the item you get to.

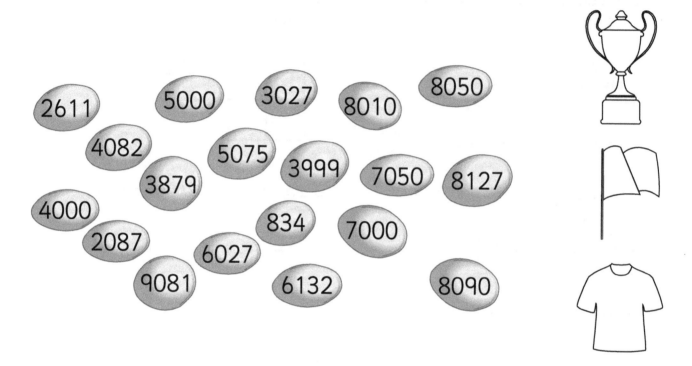

2 Write the 1st, 3rd, 8th and 10th numbers you landed on.

_____ _____ _____ _____

3 Fill in the boxes with **<** or **>** to make each statement correct.

a 8616 ☐ 8166

b 2467 ☐ 4267

c 3142 ☐ 3412

d 2655 ☐ 2650

e 5099 ☐ 9055

f 4009 ☐ 4000

Topic 5 Temperature

Reading Temperatures

1 Circle the cooler item in each pair.

a

b

2 Colour the thermometers to show the temperatures.

a

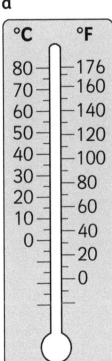

10°C = 50°F

b

15°C = 59°F

c

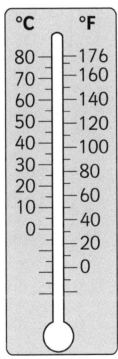

30°C = 86°F

d

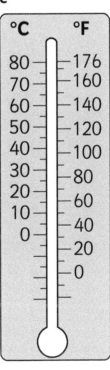 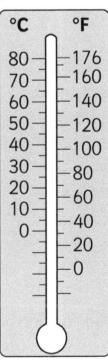

80°C = 176°F

Recording and Interpreting Temperatures

Use this chart to record your estimates and measurements.

	Place: _____ Time: _____	Place: _____ Time: _____	Place: _____ Time: _____	Place: _____ Time: _____
Estimate				
Measurement				
Difference				

What did you find out?

_____

_____

_____

Topic 6 Talking About Time

How Long Does It Take?

1 Think about these activities and events. How much time do they take?

taking a shower	getting dressed	eating a meal
playing a game of football		getting to school

a Underline the activities that take about the same amount of time.

b Write the activities in the chart. Add another activity to complete it.

Short Time	Long Time

2 Think of two activities that you do quite often. Estimate how long each activity takes. Then time yourself to see how long they really take. Work out the difference. Record your answers on the chart.

	Activity	Estimated Time	Measured Time	Difference
1				
2				

A.M. or P.M.?

Draw up a plan to show what you are going to do on your birthday.

- Think about what you will need to do to prepare: for example, blow up some balloons.

- Think about what activities you will do: for example, play a game.

Fill in the times and the activities. Use a.m. and p.m. for the times.

My birthday is on _____	
Times	**Activities**

Using a Calendar

1 Use this information to fill in the calendar.

This is the month of August. August has 31 days.
This year, the first day of August is a Tuesday.

Month: _____

Mon	Tues	Wed	Thurs	Fri	Sat	Sun

2 Complete these sentences about the calendar.

a The last day of the month is a _____.

b The date of the third Friday in the month is the _____.

c Emancipation Day is on a _____.

d There are _____ Mondays in August.

3 Your class is planning a class party on 16 December. You have to do several tasks before the party to prepare. Other tasks need to be done on the day of the party and some tasks need to be done after the party.

a Read the tasks and decide in which order you should do them. Number the tasks.

b Use the calendar and fill in the dates on which you will do each task.

December						
Mon	**Tues**	**Wed**	**Thurs**	**Fri**	**Sat**	**Sun**
			1	2	3	4
5	6	7	8	9	10	11
12	13	14	15	16	17	18
19	20	21	22	23	24	25
26	27	28	29	30	31	

Order	Task	Date
	Bake some cakes, make sandwiches.	
	Collect plates, forks and cups to use at the party.	
	Decorate the classroom.	
	Clean up the classroom and throw away all the litter.	
	Buy cold drinks.	
	Write a letter to invite the teachers.	

Topic 7 Number Facts

Addition and Subtraction Facts to 10

1 Marcie has these pieces of ribbon. Colour pairs to make lengths of 10 cm. Use a different colour for each pair.

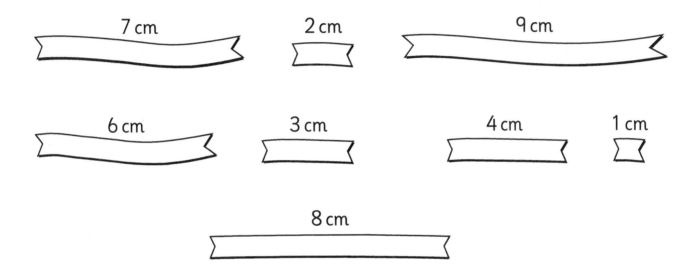

2 Fill in the missing numbers in these flow diagrams.

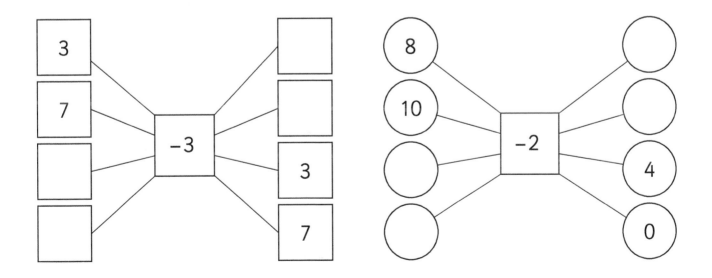

Addition and Subtraction Facts to 20

1 Pair the sets of crayons to make groups of 20. Colour each group of 20 a different colour.

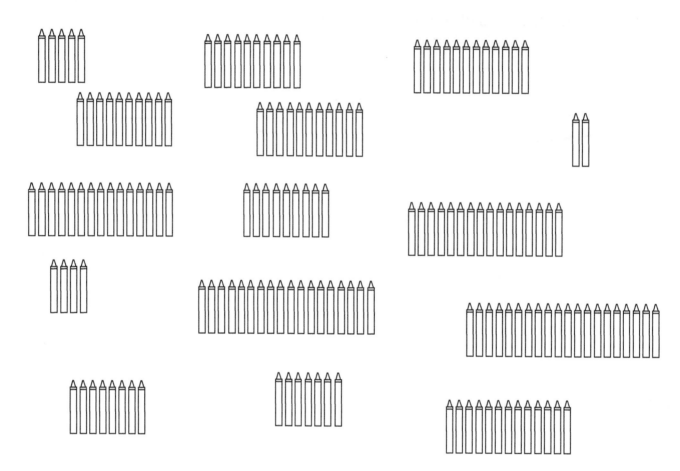

2 Use your coloured sets to complete these subtraction sentences.

a 20 – 5 = ☐

b 20 – 18 = ☐

c 20 – 0 = ☐

d 20 – 4 = ☐

e 20 – ☐ = 14

f 20 – ☐ = 9

More Adding and Subtracting

1 How quickly can you complete this set of facts?

The answer is 20.

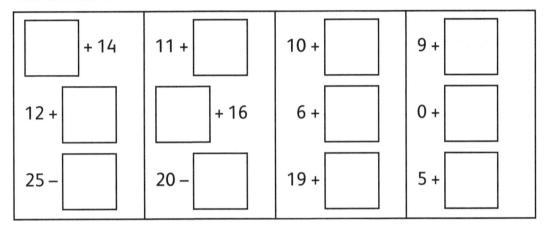

☐ + 14	11 + ☐	10 + ☐	9 + ☐
12 + ☐	☐ + 16	6 + ☐	0 + ☐
25 − ☐	20 − ☐	19 + ☐	5 + ☐

2 The number of people who get on and off a bus at different stops is given.
Work out how many people are on the bus as it leaves each stop.
Write the number on the bus.

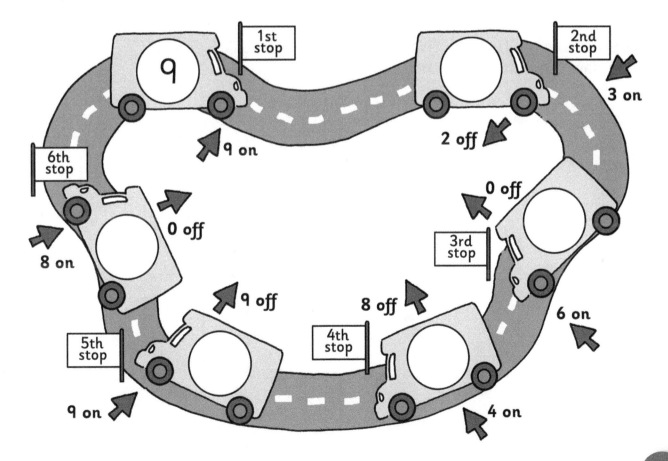

Topic 8 Classifying Shapes

Plane Shapes

1 Draw the shapes. Measure and label the sides.

Triangle	Square
Rectangle	**Pentagon**

2 Measure from the centre of this circle to points A, B, C, D, E and F.
Write down your measurements.

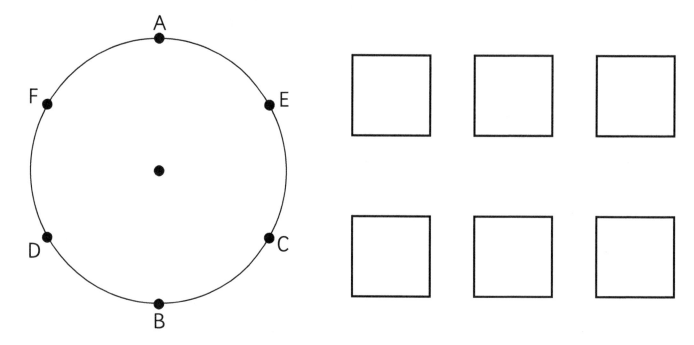

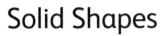

Solid Shapes

1 Complete this table.

	Name	Drawing	Number of Faces	Number of Edges
a	cube			
b			2	2
c				
d				

2 Circle the shape that does not belong in each set.

a    **b**

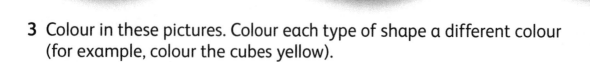

3 Colour in these pictures. Colour each type of shape a different colour (for example, colour the cubes yellow).

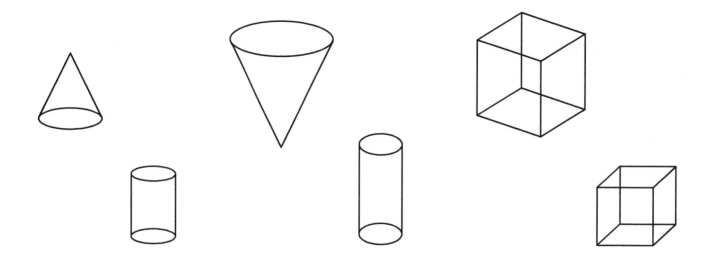

4 Make solid shapes. You can use any materials (for example, paper, clay, sticks, polystyrene).

Draw a picture of each shape you make.

Name of shape:	Name of shape:
_____	_____
What did you use to make the shape?	What did you use to make the shape?
_____	_____
Describe what the shape looks like.	Describe what the shape looks like.
_____	_____
_____	_____

Lines, Points and Line Segments

1 Draw line segments between the points.

Measure the line segments.

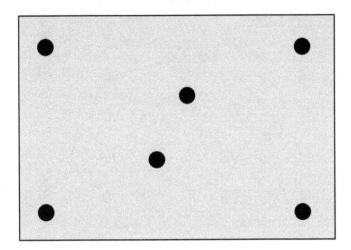

2 Add a 3 cm section and a 4 cm section to each path.

a

b

c

d

Topic 9 Rounding and Estimating

Rounding Numbers

1 For each set of numbers, round each number to the nearest ten. Write the numbers in the correct columns in the tables.

a

Rounds to 10	Rounds to 20

7 23 11
8 16
14 13 12
24 15 9

b

Rounds to 70	Rounds to 80

79
68 72
81
75 65
76 84 73

c

Rounds to 100	Rounds to 110

96 109 111
104
105 101 112 103
99 114

2 One number in each set was rounded to the nearest hundred shown above.
Colour that number.

400

398	452	349

200

147	258	199

300

193	247	269

600

713	575	650

400

491	399	458

100

107	170	201

200

127	175	250

900

870	982	950

1000

981	909	948

800

850	749	830

700

649	687	751

500

662	559	530

3 The number of pieces of litter left on a beach each day for week is recorded in the table.

Mon	Tues	Wed	Thurs	Fri	Sat	Sun
210	475	310	198	456	662	559

Round each number to the nearest hundred and estimate how much litter was left on the beach in that week.

_____ pieces of litter.

27

Estimating Answers

Round the numbers to the nearest ten or hundred. Estimate the answers.

Calculation	Rounding and Estimated Answer
a 56 + 49	
b 42 + 81	
c 48 + 33	
d 55 + 28	
e 64 + 39	
f 39 + 28	
g 66 – 19	
h 59 – 22	
i 92 – 56	
j 93 – 55	
k 97 – 79	
l 153 + 198	
m 207 + 448	
n 131 + 499	
o 688 – 420	
p 891 – 175	
q 550 – 320	

Topic 10 Number Patterns and Relationships

Odd and Even Numbers

1 Find and colour a path over the river, hopping only on even numbers.

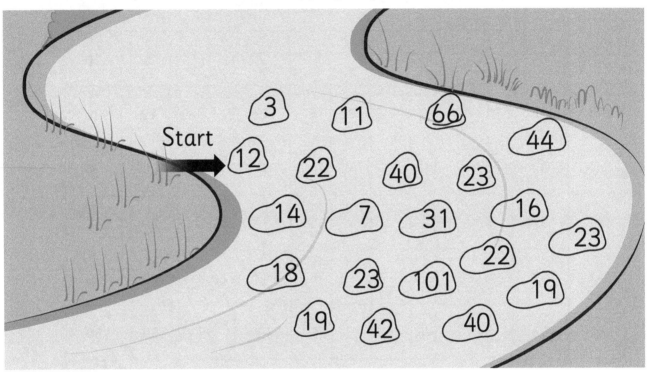

2 Circle the odd numbers. Underline the even numbers. Use the Venn diagram to organize them. Label each circle in the diagram.

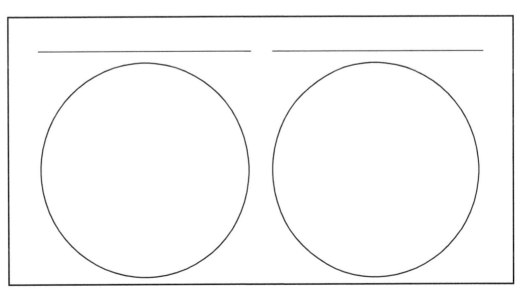

181	200
987	6
23	22
10	33
313	144
404	90
88	500
91	309
77	48

Skip-counting

1 Count by 3s. Draw a □ over each number you count.

2 Count by 4s. Draw a ○ over each number you count.

3 Count by 5s. Draw a ◇ over each number you count.

1	2	3	4	5	6	7	8	9	10
11	12	13	14	15	16	17	18	19	20
21	22	23	24	25	26	27	28	29	30
31	32	33	34	35	36	37	38	39	40
41	42	43	44	45	46	47	48	49	50
51	52	53	54	55	56	57	58	59	60
61	62	63	64	65	66	67	68	69	70
71	72	73	74	75	76	77	78	79	80
81	82	83	84	85	86	87	88	89	90
91	92	93	94	95	96	97	98	99	100

4 Choose one more skip-counting pattern. Use a triangle to mark the numbers you count.

5 Fill in the numbers from 101 to 200 on this grid.

101									
									200

a Count by 3s from 101. Colour each number you count yellow. Underline the odd numbers you count.

b Count by 5s from 105. Circle the numbers you count in blue. Underline the even numbers you count.

c Count by 50s from 100. Draw a star on each number you count.

d Count backwards by 10s from 200. Draw a red square round each number you count.

e Choose one more skip-counting pattern. Use green to show it on the grid. Describe it in words.

_____

6 Write the next three numbers in each pattern.

   a 1, 3, 5, _____, _____, _____

   b 17, 20, 23, _____, _____, _____

   c 130, 120, 110, _____, _____, _____

   d 175, 170, 165, _____, _____, _____

7 Follow the instructions. Write the numbers you would count.

   a Count by 50s. 450 ☐ ☐ ☐ ☐ ☐

   b Subtract 10. 128 ☐ ☐ ☐ ☐ ☐

   c Count back by 4s. 140 ☐ ☐ ☐ ☐ ☐

   d Subtract 2. 800 ☐ ☐ ☐ ☐ ☐

   e Subtract 50. 800 ☐ ☐ ☐ ☐ ☐

8 A packet of chips has a mass of 50 grams. Work out the mass of:

   a 5 packets _____   **b** 10 packets _____

Topic 11 Money

Working with Money

What would you have to pay for each group of items? Work out the total cost and the change.

Items	Total Cost	You Pay With	Change
$0.40 $0.65 $1.25		$5.00	
$1.20 $2.17 $1.35		$5.00	
$1.20 $2.50 $0.90 $0.90		$10.00	
$6.00 $3.50 $5.20		$15.00	
$4.75 $8.90 $3.55		$20.00	

Topic 12 Addition Strategies

Methods of Adding

Each number in the number pyramid is the sum of the two numbers in the blocks below it.

The first pyramid has been completed as an example.

Work out the missing numbers in the other pyramids. Use the space below each one for working out if you need it.

1

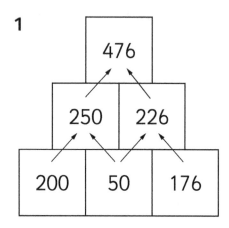

2

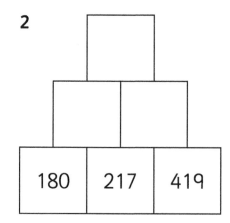

3

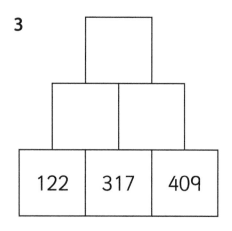

4

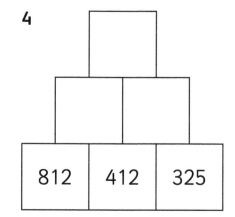

More Adding

1 Some digits are missing from these sums. Work out what they are and write them in the correct places.

a

```
  3   3   [ ]
+ 6  [ ]   8
─────────────
  9   5   5
```

b

```
      4  [ ]   7
+    [ ]   1   8
────────────────
      8   0   5
```

c

```
      9  [ ] [ ]
+   [ ]   3   4
─────────────────
[ ]   6   3   3
```

d

```
      5   1  [ ]
+   [ ]   4   7
─────────────────
  1   3  [ ]   8
```

2 Find two pairs of numbers in the box that will each give a total of 6 555 when added together. Write them in the spaces provided.

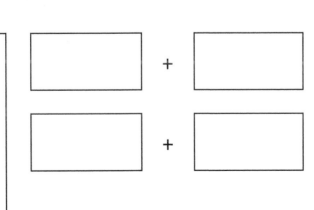

6 555		
4 234 2 411		
2 855 4 550		
3 500 3 550		
3 005 2 321		

[ ] + [ ]

[ ] + [ ]

3 Micah has found a different method of working out sums using place value.

He makes dots on the place value table to represent the numbers and then uses those to get the answer.

	Th	H	T	O
3 1 2 4	∴	•	••	∷	← 1st number

	Th	H	T	O
+ 3 1 2 5	∴	•	••	∷•	← 2nd number

	Th	H	T	O
6 2 4 9	6	2	4	9	← Totals

He has already drawn dots for the top number in each sum. Draw dots for the second number and find the answers.

a 1 2 7 8

   + 3 1 2 2

Th	H	T	O
•	∴	∷•	∷∷

b 4 0 7 2

   + 2 9 7 3

Th	H	T	O
∷		∷•	••

c 4 1 3 9

   + 2 7 8 0

Th	H	T	O
∷	•	•∷	∷∷

Topic 13 Subtraction Strategies

Methods of Subtracting

Complete these number pyramids using subtraction. Remember, each number is the sum of the two numbers below it.

1

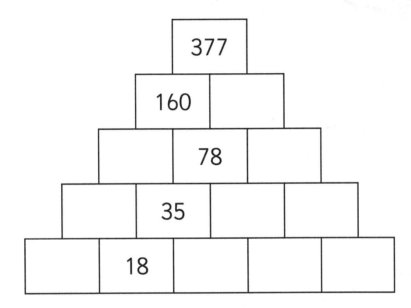

2

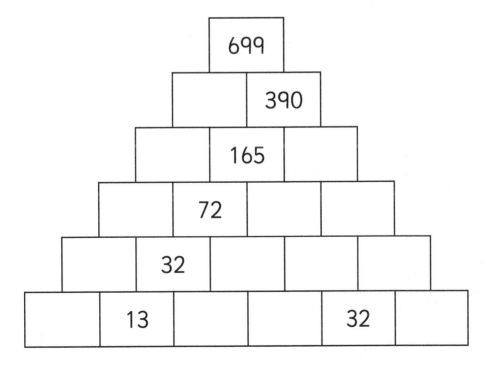

More Subtracting

1 The total distance and the distance already flown by each plane is shown on the diagrams. Work out how much further each plane has to fly.

a

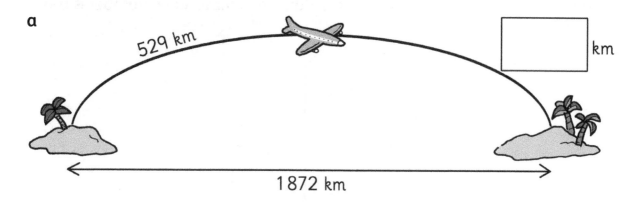

529 km

km

1872 km

b

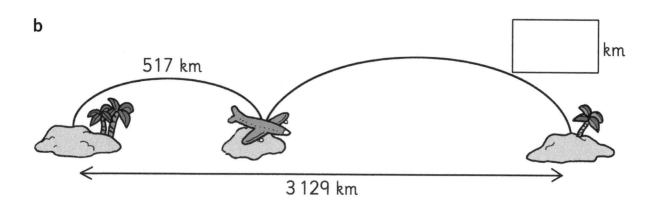

517 km

km

3 129 km

c

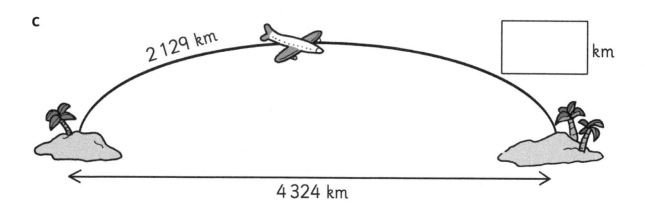

2 129 km

km

4 324 km

d

482 km

187 km

2039 km

☐ km

2 Check Tennielle's homework. If her answer is correct, tick it. If her answer is wrong, work out the correct answer and write it below the calculation.

(a)   279
     −182

    117

(b)   483
     −129

    354

(c)   673
     −308

    375

(d) 1042
     −947

    95

(e) 4200
     −1853

    2347

(f) 6000
     −2564

    4436

Topic 14 Graphs

Collecting and Recording Data

1 Work in pairs. Throw a die 20 times and record the numbers you throw on this frequency table.

Numbers	Tally	Total
1		
2		
3		
4		
5		
6		

2 Record the data on a pictograph. Choose a symbol to represent the animals.

8 rabbits   3 goats   2 ponies   10 ducks   5 hens   2 dogs

Drawing Graphs

Use this page to draw your own bar graph.

Title: _____

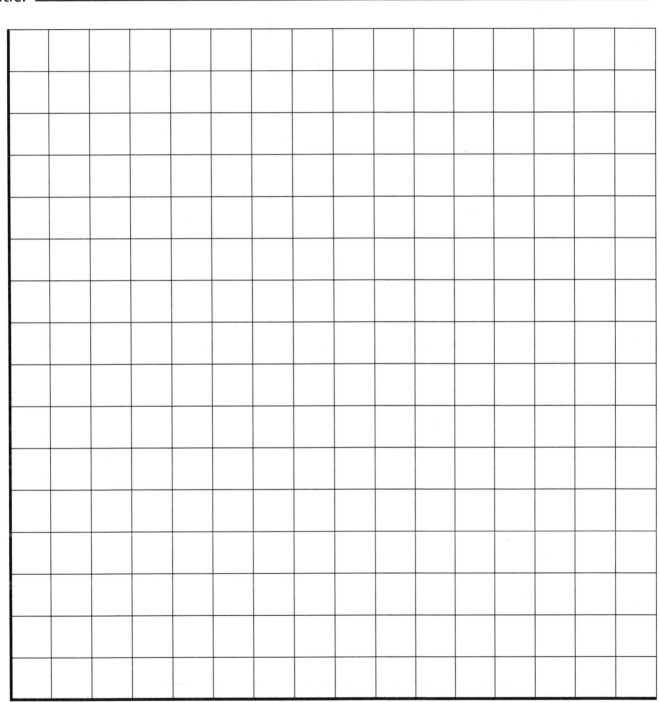

_____

Use this page to draw your own bar graph.

Title: _____

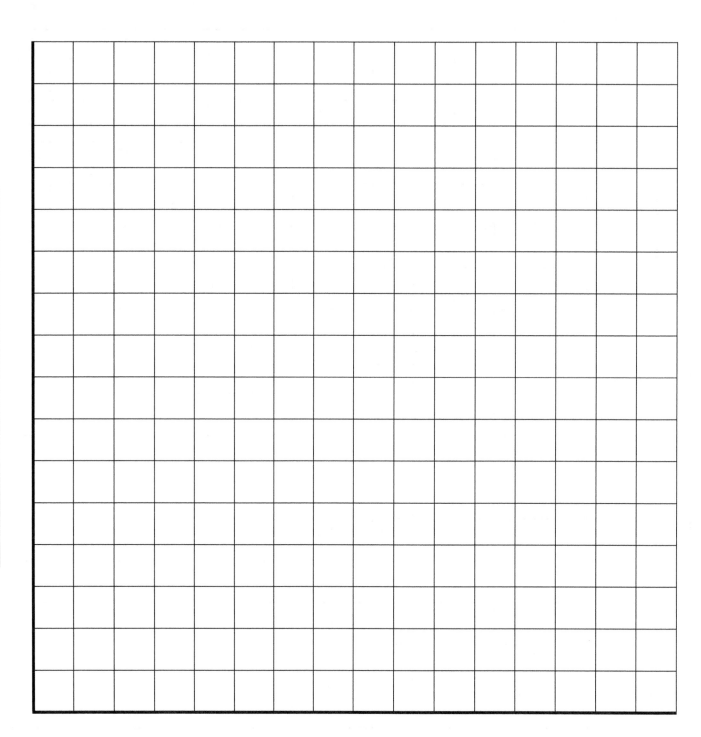

_____

1 Ernest asked everyone in his class what they liked to put in their sandwiches. He recorded the results on this tally table.

a Complete the totals.

Sandwich Filling	Tally	Total
Cheese	卌 II	7
Egg	卌	
Chicken	卌 I	
Meat	II	
Jam	卌 卌 II	

b Complete the bar graph.

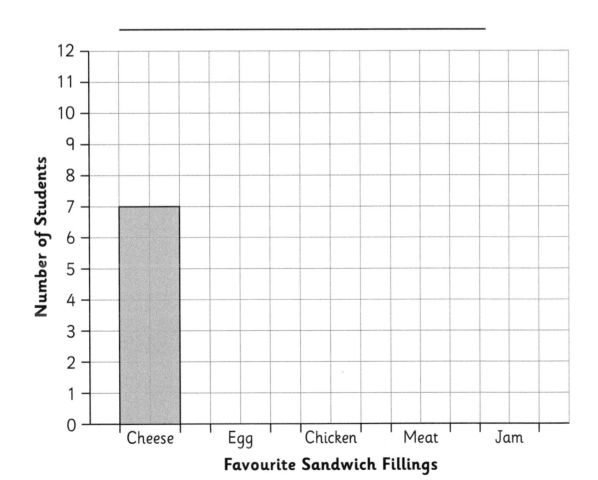

Topic 15 Problem Solving Strategies

Choosing a Strategy

1 Zorgs have 4 eyes and yabs have 2 eyes.

Draw zorgs, yabs or a combination of zorgs and yabs in each box so that there is a total of 16 eyes in every box.

zorgs   yabs

2 Put the numbers in ascending order to crack the code and work out the message.

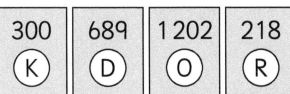

| 300 (K) | 689 (D) | 1 202 (O) | 218 (R) |

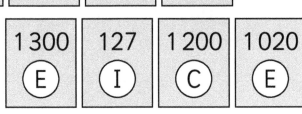

| 1 300 (E) | 127 (I) | 1 200 (C) | 1 020 (E) |

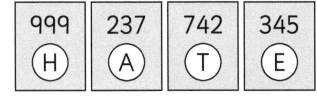

| 999 (H) | 237 (A) | 742 (T) | 345 (E) |

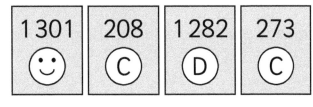

| 1 301 (☺) | 208 (C) | 1 282 (D) | 273 (C) |

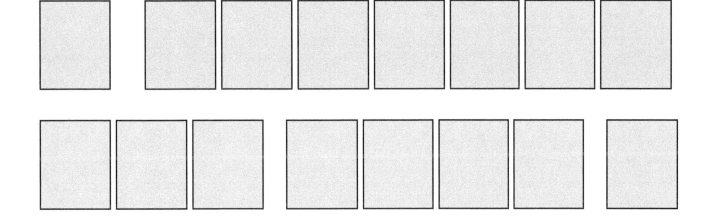

3 The picture shows the shapes you can make by joining three squares. The rule is that squares have to touch each other and when they touch, the whole side of one has to touch the whole side of the other.

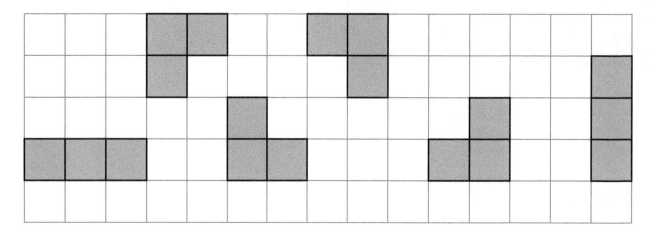

Draw all the shapes you can make with four squares using the same rule.

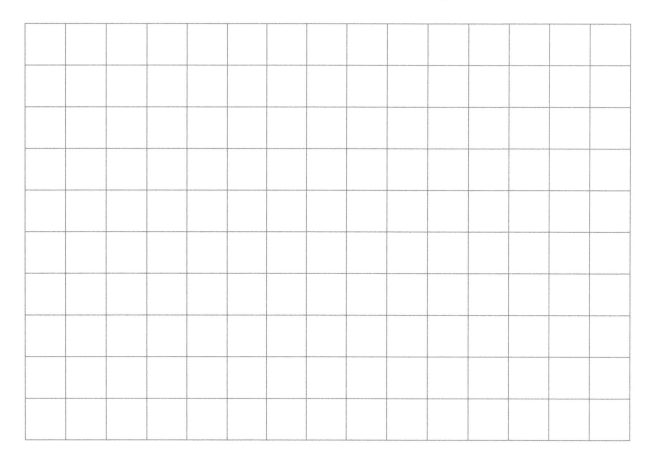

Topic 16 Measuring Length

Measuring in Metres

1 Draw pictures of two items that you would measure in metres. Explain why you would measure in metres and not in centimetres.

| | I would measure this

_____ in metres

because

_____

_____

_____

_____

_____ |
| | I would measure this

_____ in metres

because

_____

_____

_____

_____

_____ |

Measuring in Millimetres, Centimetres and Decimetres

1 Look at these objects.

   a Estimate how long the objects are in centimetres. Write your estimates in the table.

   b Now measure each object in the picture using centimetres. Write down your answers in the table.

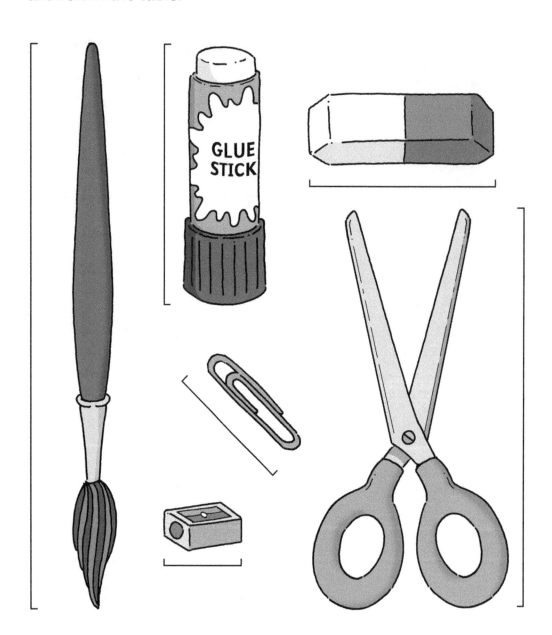

Object	Estimate (in centimetres)	Measurement (in centimetres)
Pair of scissors		

2 Measure the objects in the picture below in millimetres and centimetres. Record your results in the table.

a

b

Object	Centimetres	Millimetres
Pencil		

3 Measure these paths in centimetres or in millimetres. Write down the measurements of each section, then add them.

a

☐ + ☐ = ☐

b

☐ + ☐ = ☐

c

☐ + ☐ + ☐ = ☐

d

☐ + ☐ + ☐ + ☐ = ☐

Topic 17 Multiplying and Dividing

Multiplication and Division Facts

1 Use your class fact city posters to complete each table.

2 times table	3 times table	4 times table	5 times table
$0 \times 2 =$ ☐	$0 \times 3 =$ ☐	$0 \times 4 =$ ☐	$0 \times 5 =$ ☐
$1 \times 2 =$ ☐	$1 \times 3 =$ ☐	$1 \times 4 =$ ☐	$1 \times 5 =$ ☐
$2 \times 2 =$ ☐	$2 \times 3 =$ ☐	$2 \times 4 =$ ☐	$2 \times 5 =$ ☐
$3 \times 2 =$ ☐	$3 \times 3 =$ ☐	$3 \times 4 =$ ☐	$3 \times 5 =$ ☐
$4 \times 2 =$ ☐	$4 \times 3 =$ ☐	$4 \times 4 =$ ☐	$4 \times 5 =$ ☐
$5 \times 2 =$ ☐	$5 \times 3 =$ ☐	$5 \times 4 =$ ☐	$5 \times 5 =$ ☐
$6 \times 2 =$ ☐	$6 \times 3 =$ ☐	$6 \times 4 =$ ☐	$6 \times 5 =$ ☐
$7 \times 2 =$ ☐	$7 \times 3 =$ ☐	$7 \times 4 =$ ☐	$7 \times 5 =$ ☐
$8 \times 2 =$ ☐	$8 \times 3 =$ ☐	$8 \times 4 =$ ☐	$8 \times 5 =$ ☐
$9 \times 2 =$ ☐	$9 \times 3 =$ ☐	$9 \times 4 =$ ☐	$9 \times 5 =$ ☐
$10 \times 2 =$ ☐	$10 \times 3 =$ ☐	$10 \times 4 =$ ☐	$10 \times 5 =$ ☐

6 times table	7 times table	8 times table	9 times table
0 × 6 =	0 × 7 =	0 × 8 =	0 × 9 =
1 × 6 =	1 × 7 =	1 × 8 =	1 × 9 =
2 × 6 =	2 × 7 =	2 × 8 =	2 × 9 =
3 × 6 =	3 × 7 =	3 × 8 =	3 × 9 =
4 × 6 =	4 × 7 =	4 × 8 =	4 × 9 =
5 × 6 =	5 × 7 =	5 × 8 =	5 × 9 =
6 × 6 =	6 × 7 =	6 × 8 =	6 × 9 =
7 × 6 =	7 × 7 =	7 × 8 =	7 × 9 =
8 × 6 =	8 × 7 =	8 × 8 =	8 × 9 =
9 × 6 =	9 × 7 =	9 × 8 =	9 × 9 =
10 × 6 =	10 × 7 =	10 × 8 =	10 × 9 =

2 Complete the flow diagrams using the facts you have learned.

a
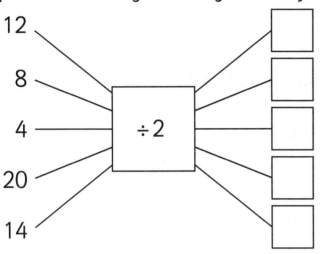

b
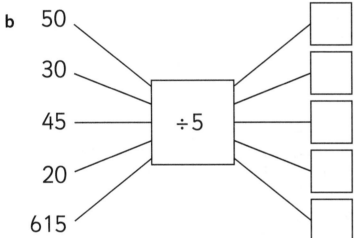

c
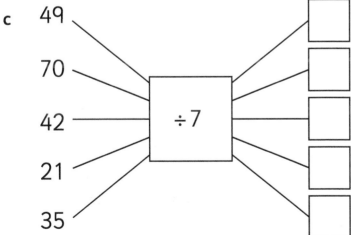

Topic 18 Fractions

Understanding Fractions

1 Shade each shape to show the fraction given. Write the fraction that is **unshaded** in the boxes.

a

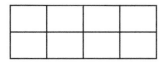

Shade $\dfrac{3}{8}$ $\dfrac{\square}{\square}$

b

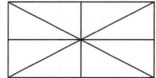

Shade $\dfrac{3}{4}$ $\dfrac{\square}{\square}$

c

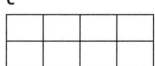

Shade $\dfrac{1}{4}$ $\dfrac{\square}{\square}$

d

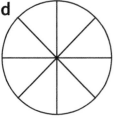

Shade $\dfrac{1}{2}$ $\dfrac{\square}{\square}$

e

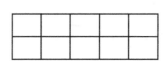

Shade $\dfrac{4}{10}$ $\dfrac{\square}{\square}$

f

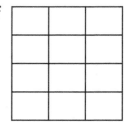

Shade $\dfrac{1}{12}$ $\dfrac{\square}{\square}$

g

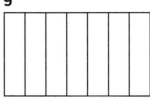

Shade $\dfrac{2}{7}$ $\dfrac{\square}{\square}$

h

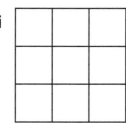

Shade $\dfrac{5}{8}$ $\dfrac{\square}{\square}$

i

Shade $\dfrac{3}{9}$ $\dfrac{\square}{\square}$

2 Colour items in the group to show the fraction given. Write a fraction to represent the unshaded items.

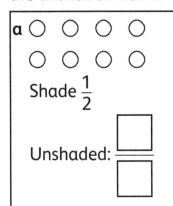

a Shade $\frac{1}{2}$

Unshaded: $\frac{\square}{\square}$

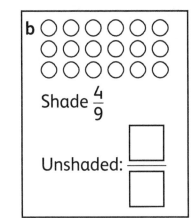

b Shade $\frac{4}{9}$

Unshaded: $\frac{\square}{\square}$

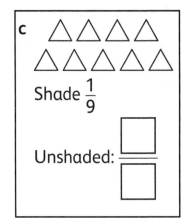

c Shade $\frac{1}{9}$

Unshaded: $\frac{\square}{\square}$

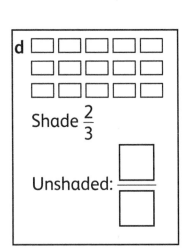

d Shade $\frac{2}{3}$

Unshaded: $\frac{\square}{\square}$

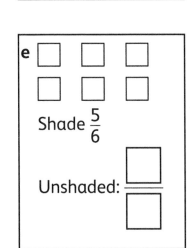

e Shade $\frac{5}{6}$

Unshaded: $\frac{\square}{\square}$

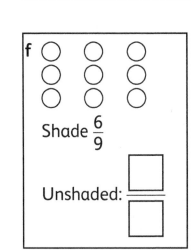

f Shade $\frac{6}{9}$

Unshaded: $\frac{\square}{\square}$

g Shade $\frac{3}{4}$

Unshaded: $\frac{\square}{\square}$

h Shade $\frac{1}{3}$

Unshaded: $\frac{\square}{\square}$

i Shade $\frac{4}{7}$

Unshaded: $\frac{\square}{\square}$

Equivalent Fractions

1 Colour $\frac{1}{2}$ of each shape. Complete the equations.

a

$$\frac{1}{2} = \frac{\square}{4}$$

b

$$\frac{1}{2} = \frac{3}{\square}$$

c
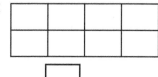

$$\frac{1}{2} = \frac{\square}{8}$$

d
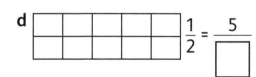

$$\frac{1}{2} = \frac{5}{\square}$$

e

$$\frac{1}{2} = \frac{\square}{\square}$$

2 Colour $\frac{3}{4}$ of each shape. Complete the equations.

a
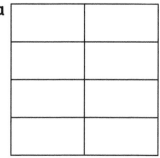

$$\frac{3}{4} = \frac{\square}{8}$$

b
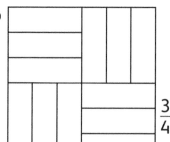

$$\frac{3}{4} = \frac{\square}{\square}$$

3 Colour $\frac{2}{3}$ of each shape. Complete the equations.

a
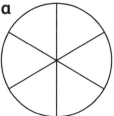

$$\frac{2}{3} = \frac{\square}{6}$$

b
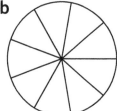

$$\frac{2}{3} = \frac{\square}{9}$$

c
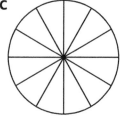

$$\frac{2}{3} = \frac{\square}{\square}$$

More Equivalent Fractions

1 Draw lines between pairs of equivalent fractions. Colour any fractions that are in simplest form.

$\dfrac{3}{4}$

$\dfrac{4}{5}$

$\dfrac{11}{33}$

$\dfrac{1}{2}$

$\dfrac{14}{28}$

$\dfrac{16}{20}$

$\dfrac{27}{30}$

$\dfrac{18}{24}$

$\dfrac{9}{10}$

$\dfrac{14}{21}$

$\dfrac{9}{15}$

$\dfrac{1}{3}$

$\dfrac{4}{10}$

$\dfrac{2}{5}$

$\dfrac{3}{5}$

$\dfrac{2}{3}$

Topic 19 Working with Time

Talking Time

Sort these activities into a.m. and p.m. activities. List the activities in the correct column.

a a birthday party which starts at 10:00

b a sun which sets at 8:30

c a trip to the shop at 11:15

d eating breakfast

e having a shower before school

f playing with friends after school

g doing homework after school

h eating supper

a.m. Activities	p.m. Activities

How Long Does It Take?

1 Complete all the columns in the table.

Activity	Start	End	How Long Did It Take?
Making a sandwich			15 minutes
Walking to the beach			
			30 minutes
Making a cake			

2 Sam takes 10 minutes to walk to the beach, 30 minutes to play cricket with his friends, 10 minutes to look for shells and 5 minutes to buy a cold drink. He then walks home again. How long has he been out altogether?

3 Fill in the calendar. The month is September. The month starts on a Wednesday.

_____

Mon						

4 Use your calendar to answer these questions.

   a How many Saturdays are there in September? _____

   b You have music lessons twice a week, on Tuesdays and Thursdays.
     How many music lessons do you have in September?

   _____

   c There is a carnival on the 3rd weekend in the month. What are the dates
     of the carnival?

   _____

   d Sean's birthday is on the fourth Wednesday of September. What is the date?

   _____

   e Your father is going on a trip from the 6th to the 14th of September.
     How many days will he be away?

   _____

Equivalent Times

1 Match the times. Write the pairs below.

(3 hours)	24 hours	$3\frac{1}{2}$ days	730 days	3 weeks
	36 months	300 minutes		6 weeks
3 years	120 seconds		21 days	
		84 hours		(180 minutes)
	24 months		5 hours	
1 day	2 minutes			
42 days		49 days		7 weeks

a 3 hours _____ = 180 minutes _____

b _____ = _____

c _____ = _____

d _____ = _____

e _____ = _____

f _____ = _____

g _____ = _____

h _____ = _____

i _____ = _____

j _____ = _____

Topic 20 Decimal Fractions

Tenths and Hundredths in Decimal Form

1 Shade the given fractions. Write each one as a decimal.

a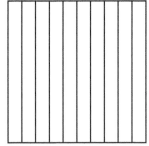

$\dfrac{2}{10}$ = 0. _____

b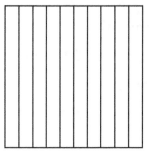

$\dfrac{5}{10}$ = 0. _____

c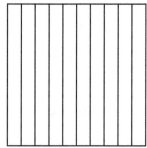

$\dfrac{9}{10}$ = 0. _____

d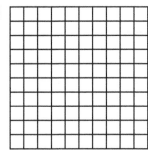

$\dfrac{3}{10}$ = 0. _____

e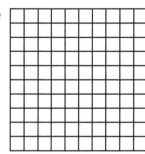

$\dfrac{1}{10}$ = 0. _____

f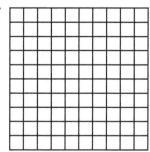

$\dfrac{7}{10}$ = 0. _____

2 Shade $\dfrac{1}{2}$ of the first grid and $\dfrac{3}{4}$ of the second grid.

Write the shaded section of each as a decimal fraction.

a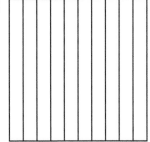

0. _____ shaded

b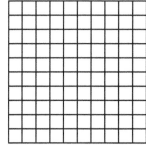

0. _____ shaded

3 Colour part of each hundred square to match the given decimal fraction.
Complete the number sentence.

a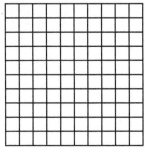

0.09

0.09 + 0. _____ = 1

b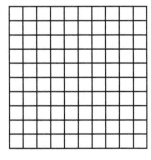

0.90

0.90 + 0. _____ = 1

c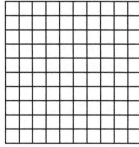

0.55

0.55 + 0. _____ = 1

d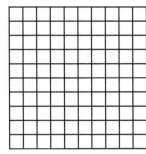

0.75

0.75 + 0. _____ = 1

e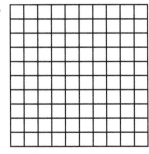

0.99

0.99 + 0. _____ = 1

f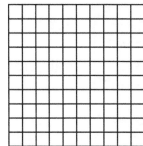

0.8

0.8 + 0. _____ = 1

4 Fill in the boxes with **>**, **=** or **<** to make each statement correct.

a $\frac{3}{10}$ ☐ 0.3

b 0.26 ☐ 0.22

c 1 ☐ 0.99

d 0.32 ☐ 0.3

e 0.45 ☐ 0.5

f 0.5 ☐ 0.1

Topic 21 Capacity and Mass

Measuring Capacity

1 Which unit would you use to measure the capacity of each object?
Circle the correct answer.

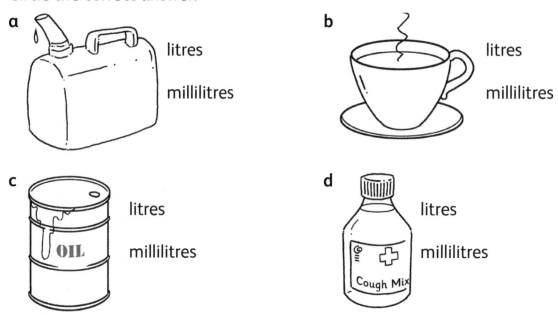

a

litres

millilitres

b

litres

millilitres

c

litres

millilitres

d

litres

millilitres

2 Which container in each pair has the greater capacity? Circle the pictures.

a

b

c

d

Measuring Mass

1 Estimate the mass of each object. Circle the mass that is closer.

a  4 kg

400 g

b   1 kg

100 g

c   200 g

2 kg

d  25 kg

250 g

e  250 g

2.5 kg

2 Use this chart to record measurements of mass.

	Object	Estimate of Mass	Actual Mass	Difference
1				
2				
3				
4				
5				

Topic 22 Exploring Shapes

Lines of Symmetry

1 Circle the objects that are symmetrical.

2 Draw lines of symmetry through these shapes. Both sides must be exactly the same.

3 Complete these shapes to make them symmetrical.

a
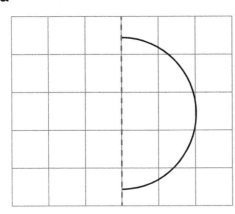

b
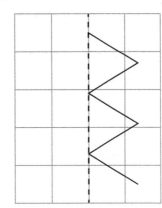

c
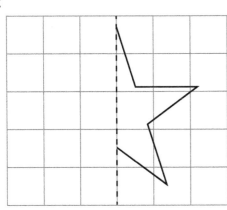

d
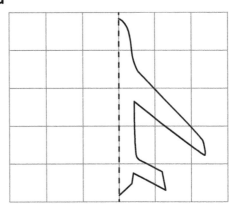

e
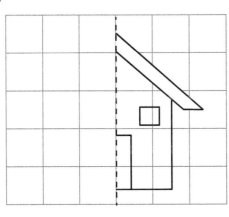

f
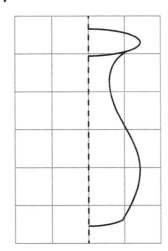

Moving Shapes

1 a Draw four small, simple shapes on coloured paper. The shapes should have straight and curved lines. Cut out your shapes.

b Trace each shape in the first column of the table.

c Move the shapes. Slide them, flip them and turn them. Complete the table by drawing the shapes to show their positions when you move them.

Shape	Slide the Shape	Flip the Shape	Turn the Shape

Topic 23 Perimeter and Area

Perimeter

1 Measure and write down the perimeter of each shape.

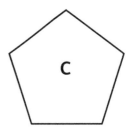

A

B

C

_____

_____

_____

D

E

F

_____

_____

_____

2 Which shape has the greatest perimeter? Write them down in order from the greatest to the least.

Shape						
Perimeter						

3 Draw two different shapes, A and B, each with a perimeter of 15 cm.
Measure and label the sides of your shapes.

A

B

Area

1 Find the area of each shape. ☐ = one square centimetre

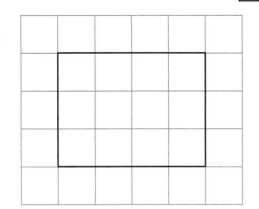

A = _____ square cm

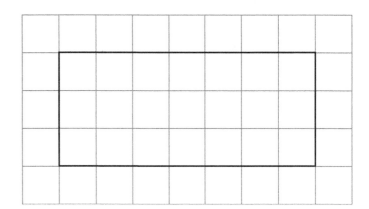

B = _____ square cm

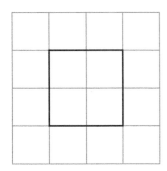

C = _____ square cm

D = _____ square cm

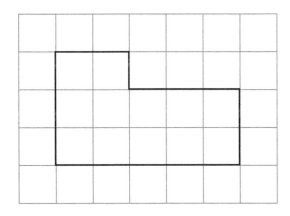

E = _____ square cm

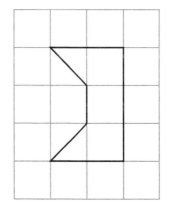

F = _____ square cm

2 Draw the following figures on this squared grid.

a a shape with an area of 12 square cm

b a shape with an area of 18 square cm

c a small animal with an area of 20–26 square cm

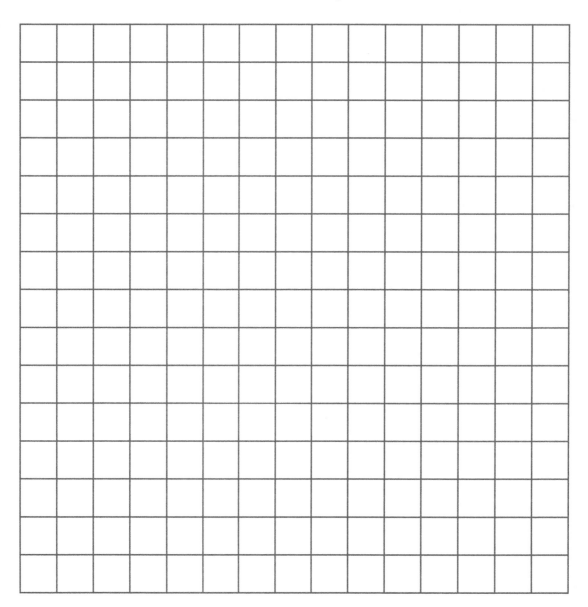

Topic 24 Probability

Certain, Possible and Impossible Events

1 Colour the counters in each bag so that the statements are true.

a

It is possible to get red, but impossible to get yellow.

b

It is possible to get two red or two yellow.

c

You are certain to get red or yellow.

d

It is possible to pull out four different colours, but impossible to get yellow.

e

You are certain to get green or blue.

f

It is impossible to get red or yellow.

2 Read the information, then colour the crayons.

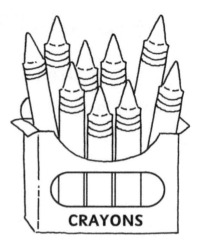

- It is possible to choose a red, blue or yellow crayon.

- It is impossible to choose a white or green crayon.

- You are certain to choose a colour with an 'e' in its name.

3 Use the calendar for this month. Write three events in each column of the chart.

This month, these events are:		
impossible	certain	possible

Topic 25 Mixed Measures

Working with Different Units

1 Draw lines to join the equivalent measurements. The first one has been done for you.

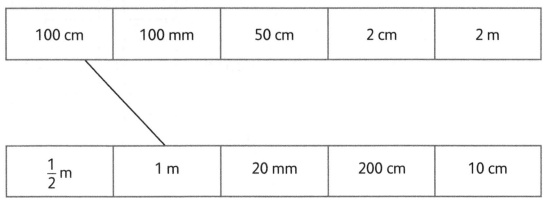

100 cm	100 mm	50 cm	2 cm	2 m

$\frac{1}{2}$ m	1 m	20 mm	200 cm	10 cm

2 Fill in the missing measurements.

a 1 cm = ☐ mm

b 1 m = ☐ cm = ☐ mm

c 1 ft = ☐ in

d ☐ ft = 1 yard

3 Write each of these measurements in centimetres.

a 4 m = ☐ cm

b 12 m = ☐ cm

c 140 mm = ☐ cm

d 1 m and 3 cm = ☐ cm

e 85 mm = ☐ cm

f 9 m and 25 cm = ☐ cm

Equivalent Measurements

1 Complete the number sentences using equivalent measurements.

a 200 cm = ☐ m

b 380 cm = ☐ m ☐ cm

c 201 cm = ☐ m ☐ cm

d 525 cm = ☐ m ☐ cm

e 1000 cm = ☐ m

f 300 mm = ☐ cm

g 2450 cm = ☐ m ☐ cm

h $\frac{1}{2}$ cm = ☐ mm

2 Karen walked 20 metres to Clare's house, then turned around and walked 50 metres in the other direction to John's house.

a How many centimetres did she walk altogether?

_____

b How many centimetres was she from her original starting point?

_____

3 Some units are missing from these equivalent measures.
Fill in the missing units.

a 316 ☐ = 3 m and 16 cm

b 316 ☐ = 31 cm and 6 mm

c 316 ☐ = 3 160 mm

d 316 ☐ = 3 160 cm

Find the Mistakes!

There is a mistake in each of these calculations. Find the mistakes and write the correct answers below each calculation.

(1) 2 m 50 cm
  + 4 m 50 cm   ✗

6 m 00 cm
_____

(2) 4 m 19 cm
  + 2 m 25 cm   ✗

6 m 39 cm
_____

(3) 3 m 35 cm
  + 4 m 80 cm   ✗

7 m 5 cm
_____

(4) 3 m 90 cm
  + 2 m 9 cm   ✗

6 m 80 cm
_____

(5) 4 m 30 cm
  − 2 m 15 cm   ✗

1 m 15 cm
_____

(6) 6 m 20 cm
  − 4 m 15 cm   ✗

2 m 35 cm
_____

(7) 4 m
  − 2 m 50 cm   ✗

2 m 50 cm
_____

(8) 10 m 30 cm
  − 7 m 40 cm   ✗

3 m 90 cm
_____

Hachette UK's policy is to use papers that are natural, renewable and recyclable products and made from wood grown in well-managed forests and other controlled sources. The logging and manufacturing processes are expected to conform to the environmental regulations of the country of origin.

Orders: please contact Hachette UK Distribution, Hely Hutchinson Centre, Milton Road, Didcot, Oxfordshire, OX11 7HH. Telephone: +44 (0)1235 827827. Email: education@hachette.co.uk. Lines are open from 9 a.m. to 5 p.m., Monday to Friday. You can also order through our website: www.hoddereducation.com

ISBN: **978 1 4718 6462 9**

© Cloud Publishing Services 2016

First published in 2016 by
Hodder Education,
An Hachette UK Company
Carmelite House
50 Victoria Embankment
London EC4Y 0DZ

www.hoddereducation.com

Impression number   10 9 8 7 6 5

Year   2023

Cover photo © Giovanni Costa/123RF.com

Illustrations by Oxford Illustrators

Typeset in India by Aptara Inc.

Printed in the UK

A catalogue record for this title is available from the British Library.